This book belongs to

Please email me if you find this notebook

info@ashleysnotebooks.com
www.ashleysnotebooks.com

www.ingramcontent.com/pod-product-compliance
Lightning Source LLC
Chambersburg PA
CBHW081010170526
45158CB00010B/2993
* 9 7 8 1 0 7 6 7 4 7 4 0 2 *